TRANSMISSION OF PRESSURE IN LIQUIDS

And

PASCAL'S PRINCIPLE

© sciencebod 2017

Preface

The principles behind transmission of pressure in liquids form the bedrock of hydraulic engineering and a whole lot of fluid-related studies. Pascal's principle is at the center of these principles, and we think it is important for new physics students in secondary schools to understand these ideas. We therefore present the ideas in a manner that makes them interesting and appealing to lean in this little book. We also encourage the students to try to solve the problems presented in this book on their own, and to convince themselves that they understand the principles as they progress through the book.

© sciencebod 2017

TRANSMISSION OF PRESSURE IN LIQUIDS AND PASCAL'S PRINCIPLE

In an attempt to perforate a cardboard sheet, a student found out that more force was required to perforate the cardboard sheet using a metal rod with a round end than using a nail of the same size as the metal rod.

The difference in the difficulty is due to pressure difference. What is pressure?

Definition.

1

Pressure is defined as the perpendicular force per unit area acting on a surface.

i.e. $\quad \text{Pressure (P)} = \dfrac{\text{Perpendicular force}(F)}{\text{Area}(A)} = \dfrac{F}{A}$ \hfill (1)

The S.I unit of pressure is $\dfrac{N}{m^2}$.

Other units of pressure are

The pascal (pa) $1 \text{ pa} = 1 Nm^{-2}$ and

The bar (bar) $\quad 1 \text{ bar} = 10^5 Nm^{-2} = 10^5 \text{pa}$.

Pressure is a scalar quantity.

Note!

2

From the definition of pressure above, we can see that pressure is inversely proportional to the perpendicular area; that is, the lesser the area, the more the pressure exerted by a constant force. A force applied on a small perpendicular area produces more pressure than the same force applied on a large perpendicular area.

This explains the difference in the pressure applied by the student who tried to perforate the cardboard sheet using a nail (pointed end) and a rod (round end). The nail has a smaller area (at the pointed end) than the metal rod (at the round end), and so if the student applies the same force on both the nail and the rod, he will get more pressure with the nail than with the rod.

It also explains why less force is required to cut an object with a sharp knife than with a blunt knife.

Let's illustrate the pressure definition with the following JAMB question (2004)

3

A man exerts a pressure of $2.8 \times 10^{-2} Nm^{-2}$ of his feet in contact with the ground over an area $4 \times 10^{-2} m^2$. The weight of the man is
(A) 102N (B) 70N (C) 100N (D) 112N

Solution!

4

$$\text{Pressure} = \frac{Force}{Area}$$

Where the man's weight is the same as the Force exerted.

⇨ Force = Pressure × Area

$$= 2.8 \times 10^3 \times 4 \times 10^{-2}$$
$$= 112N$$

So option D is correct.

And you should be able to answer correctly the following WAEC question (2001)

5

Which of the following statements about pressure is not correct?
Pressure
        (A) increases with an increase in surface area
        (B) decreases with an increase in surface area

(C) increases with a decrease in surface area

(D) increases with an increase in the applied force

Answer:

Obviously, as explained in Plan 2, Pressure decreases (not increase) when the surface area is increased. The statement in option (A) is therefore not correct, and that is the answer.

Another WAEC question (2008)

A rectangular block of dimensions 2.0m×1.0m×0.5m weighs 200N. Calculate the maximum pressure exerted by the block on a horizontal floor

(A) $100Nm^{-2}$ (B) $200Nm^{-2}$ (C) $300Nm^{-2}$ (D) $400Nm^{-2}$

Solution!

The question is a bit tricky!

This is because we have to first decide which face of the rectangular block has the smallest area. The maximum pressure will be exerted by the block when it lies (or when it touches the horizontal ground) on the face that has the smallest area.

The two smallest dimensions of the block are 0.5m and 1.0m, therefore they represent the face of the block with the smallest area. The area of this face is 0.5m × 1.0m = $0.5m^2$.

The maximum pressure that will be exerted by the block on a horizontal floor is therefore

$$\text{Pressure} = \frac{Force}{Area} = \frac{200N}{0.5m^2} = 400Nm^{-2}$$

And finally, we consider the following JAMB question (2003)

8

The stylus of a phonograph record exerts a force of 7.7×10^{-2} N on a groove of radius 10^{-5} m. Compute the pressure exerted by the stylus on the groove.

(A) $2.45 \times 10^8 Nm^{-2}$ (B) $3.45 \times 10^8 Nm^{-2}$ (C) $4.90 \times 10^8 N^{-2}$ (D) $2.42 \times 10^8 Nm^{-2}$

Solution!

9

Area of the groove = πr^2
$= 3.142 \times (10^{-5} m)^2$
$= 3.142 \times 10^{-10} m^2$

And so, the pressure exerted by the stylus on the groove is

$$P = \frac{F}{A} = \frac{7.7 \times 10^{-2} N}{3.142 \times 10^{-10} m^2}$$

$$= 2.45 \times 10^8 Nm^{-2}$$

Pressure in liquids

10

Let us assume a column of liquid of height h and base area A as illustrated in Figure 1.

The volume (V) of the liquid is given by base area (A) multiplied by the height (h) of the liquid column, that is, V = Ah.

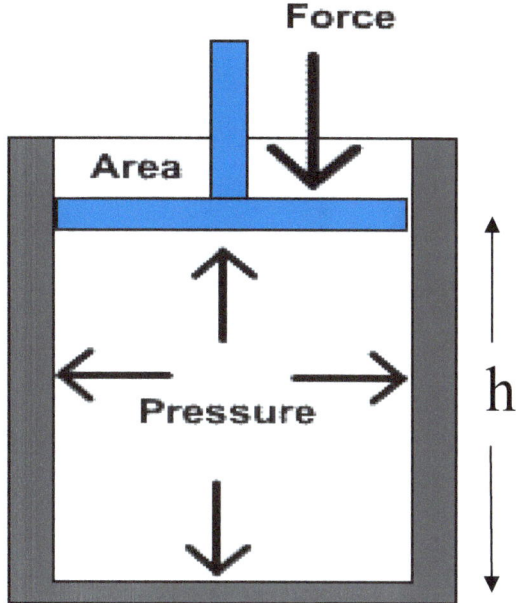

Figure 1: Pressure in Liquids (GlobalSpec, 2011)

If ρ = density of the liquid, then mass of the liquid is given by $V \times \rho = Ah \times \rho$
And weight of the liquid column is $w = m \times g = Ah \times \rho \times g$

The weight of the liquid exerts a force on the base area of the cylinder A. Therefore, the pressure at the base of the liquid column is given by

$$\text{Pressure, P} = \frac{\text{Normal force}}{\text{Area}} = \frac{W}{A} = \frac{Ah\rho g}{A} = h\rho g$$

Where h = height of the liquid column
    ρ = density of the liquid, and
    g = acceleration due to gravity

Observe that the area (A) cancelled out, so the pressure in a liquid DOES NOT depend on the cross sectional area of the container.

Characteristics of pressure in liquids

11

From the foregoing, we can summarize characteristics of pressure in liquids as follows:

(i) The pressure in a liquid increases in direct proportion to the depth of the liquid

(ii) The pressure in different liquids at the same depth varies directly with density

(iii) The pressure at any point in the liquid acts equally in all directions

(iv) The pressure at all points at the same level within a liquid is the same

(v) The pressure in a liquid does not depend on the cross sectional area of the container

And we continue with more questions to illustrate! JAMB (1993)

12

A weightless vessel of dimensions 4m × 3m × 2m is filled with a liquid of density 1000kgm^{-3} and scaled. What is the maximum pressure this container can exert on a flat surface?

(A) $9\times10^4 Nm^{-2}$ (B) $4\times10^4 Nm^{-2}$ (C) $3\times10^4 Nm^{-2}$ (D) $2\times10^4 Nm^{-2}$

Solution

13

The volume of the liquid is V = 4m × 3m × 2m = 24m^3

Density $\rho = 1000 kgm^{-3}$

And so the mass of liquid that fills the container is

m = Volume × Density
 = 24m^3 × 1000kgm^{-3}
 = 24000kg

As before, the surface of the container that gives the maximum pressure is the one with the smallest area. That is the surface with dimensions 3m × 2m
The area of this surface is, A = 3m × 2m = 6m²

And so the maximum pressure this container can exert on a flat surface is

$$\text{Pressure} = \frac{Force}{Area} = \frac{mg}{A} = \frac{24000kg \times 10ms^{-2}}{6m^2} = 4 \times 10^4 Nm^{-2}$$

Option B is therefore the right answer.

JAMB (1994)

14

Which of the following statements are true of pressure in liquids?
- (I)   Pressure acts equally in all directions
- (II) Pressure decreases with depth
- (III) Pressure at the same level of a liquid is the same
- (IV) Pressure is dependent on the cross-sectional area of the barometer tube

(A) I and III only   (B) I, II and III only   (C) I, III and IV only   (D) I, II, III and IV

Answer: Only options I and III are true, so option (A) is correct!

JAMB (1995)

15

The volume of an air bubble increases from the bottom to the top of a lake at constant temperature because

(A) Atmospheric pressure acts on the surface of the lake
(B) Pressure increases with depth.
(C) Density remains constant with pressure
(D) The bubble experiences an upthrust

Option (B) is correct. As an air bubble moves from the bottom of a lake to the top, the bubble experiences lesser pressure (this is because pressure increases with depth, and so the pressure at the top is less than the pressure at the bottom). The air bubble therefore expands because it encounters lesser pressure.

JAMB (2000)

16

At a fixed point below a liquid surface, the pressure downward is P_1 and the pressure upward is P_2. It can be deduced that

(A) $P_1 > P_2$ (B) $P_1 \geq P_2$ (C) $P_1 < P_2$ (D) $P_1 = P_2$

Solution

17

Option (A) is correct. Since point P_2 is upward of P_1, it follows that the pressure at point P_1 should be greater than at point P_2. This is obviously because the pressure in a liquid increases with depth.

Atmospheric pressure

18

Gases also exert amounts of pressure in a manner that is very similar to liquids!

The atmosphere which consists of a mixture of gases, some dust particles and water vapor, exerts some pressure on the surface of the earth. This is because its air has some weight. The pressure due to the atmosphere is called the atmospheric pressure.

We do not feel this atmospheric pressure because the pressure inside our body is equal to the external pressure. If an empty can is evacuated, it collapses because the atmospheric pressure is now much greater than the pressure inside the can.

Plan 19

19

At sea level, the atmospheric pressure is about 10^5 Pascal's. This amount of pressure is called 1 atmosphere.

The atmosphere supports a column of mercury of about 760mm high. So, one atmosphere is equal to 760mm mercury (i.e. 760mmHg).

The atmospheric pressure decreases with height or attitude. This is because at the bottom of the atmosphere, there is more column of air above, and if we go higher, the column of air above decreases. The air is also found to be denser at sea level than at higher altitudes.

Measurement of Atmospheric and Gas Pressure

20

The atmospheric and gas pressure can be measured using a variety of measuring instruments, namely,
- (i) Simple mercury barometer
- (ii) Aneroid barometer and
- (iii) Manometer

We also have others like the Forte's barometer but for the scope of this book, we shall be interested in the 3 listed above.

The simple mercury barometer

21

The simple mercury barometer is a very simple instrument used in measuring atmospheric pressure. In figure 2, a glass tube about 1 meter long is filled with mercury, making sure there are no air bubbles. It is closed with the finger, inverted into a bowl of mercury and the finger removed.

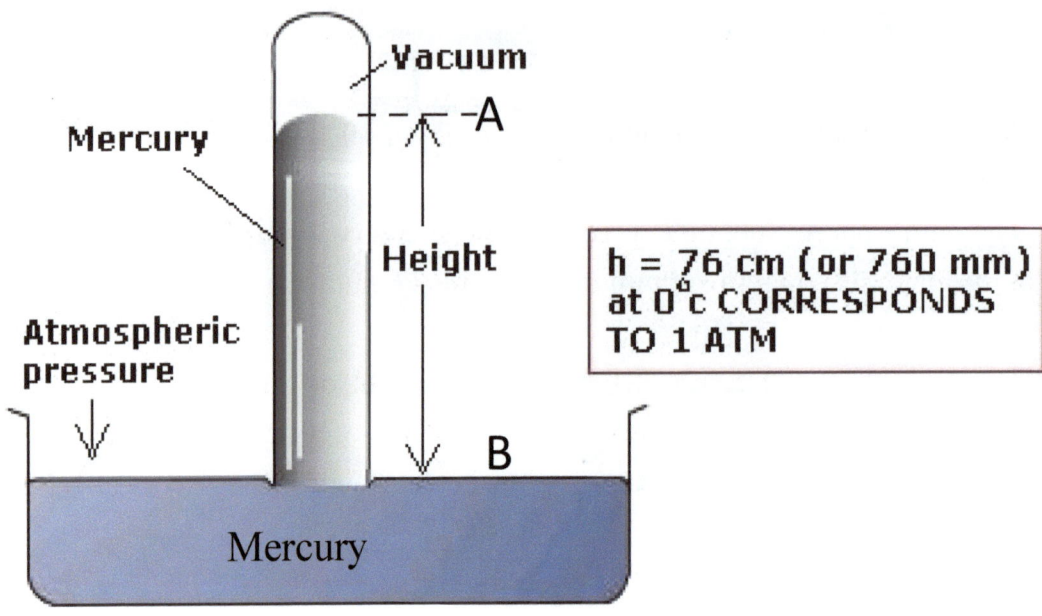

Figure 2. The simple mercury barometer (Tutorvista, 2017)

Observation!

It is usually observed that the mercury level in the glass tube falls until it is about 76cm or 760mm above the level of the mercury in the bowl shown in Figure 2. The atmospheric pressure is equal to the pressure of the column of mercury AB. This is the height of mercury which the atmosphere can support. It is also observed that the vertical height of mercury remains constant even when the tube is tilted.

The value of the atmospheric pressure is therefore equal to the pressure of 760 mm column of Mercury, and it is usually expressed as 760 mmHg. (Hg is the chemical symbol for mercury). We compute the value of this pressure using the formula:

Pressure = hρg

Where h is height of the mercury = 0.76 m,

ρ is density of mercury = 13600 kgm^{-3}

and g is the value of acceleration due to gravity = 10 ms^{-2}

Therefore:

Atmospheric pressure = Pressure of 760 mm column of Hg

$$= h\rho g$$
$$= 0.76m \times 13600 kgm^{-3} \times 10ms^{-2}$$
$$= 103360 \; kgm^{-1}s^{-2} \text{ which is same as } 103360 \; Nm^{-2} \text{ or } 103360 \; Pa$$

Defects of a simple barometer

The simple barometers are not very accurate due to the following defects:

(i) The space above the mercury is not pure vacuum. This is because there is always some air and water vapor in the space above the mercury which has the effect of making the barometer read less than the true atmospheric pressure.

(ii) The height of mercury level h is only a rough estimate since there is no provision for accurate measurement.

The Aneroid barometer

A type of barometer which works without mercury or liquid is the Aneroid barometer shown in figure 3. The aneroid barometer consists of a partially evacuated, thin metal box with a system of levers. Changes in atmospheric pressure causes the thin sides of the partially evacuated metal box to move in or out. This movement is transmitted by means of a metal spring and system of levers to a pointer which moves over a graduated scale.

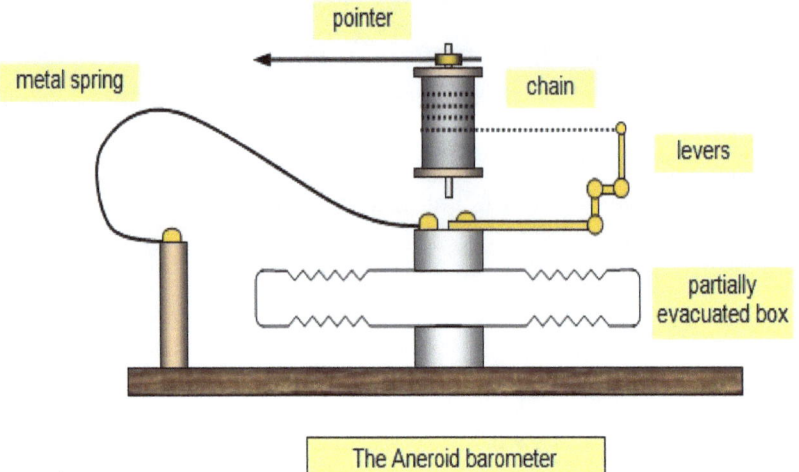
Figure 3. The Aneroid Barometer (SchoolPhysics, 2016)

Note!

One important use of a barometer is to determine altitude (i.e. as an altimeter). Originally, the barometer measures atmospheric pressure. It can however also be used to measure altitude. This is simply based on the fact that the atmospheric pressure changes with altitude (The greater the altitude, the lower the atmospheric pressure). So, a barometer can also be supplied with a nonlinear scale to indicate altitude, and when this is done the instrument is called a pressure altimeter or barometric altimeter.

The Aneroid Barometer is specifically suitable for use as an Altimeter (especially in aircrafts and other high flying experiments). This is because it has a small size and very convenient to carry without the worry of spilling any liquid. The aneroid barometer is also highly accurate if well calibrated.

The Manometer

The basics of the manometer are illustrated in Figure 4. The Manometer consists of a small bore U-tube with enlarged ends on which atmospheric pressure is acting. The liquid inside the U-tube could be water, mercury or any other suitable liquid. For simplicity, let's assume the liquid here to be water. The water level is the same when both ends of the U-tube are open to the atmosphere. This is because the same amount of pressure (the atmospheric pressure) acts on both ends.

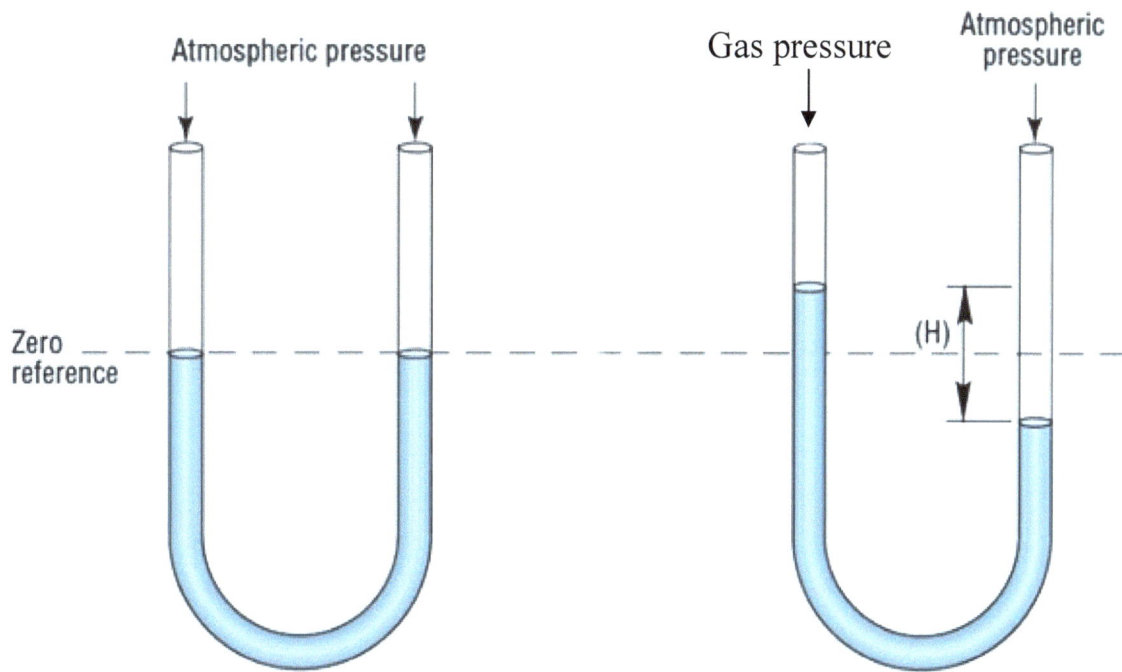

Figure 4. The Manometer (Dojo University, 2017)

The manometer can be used to measure the pressure of gas by attaching and sealing the gas to one end of the U-tube. When this is done, the water level is altered:
1. If the gas pressure is greater than the atmospheric pressure, then the gas pushes the water so that the water level in the open arm is higher than in the sealed gas arm. In this case, the gas pressure is computed as:
Gas pressure = atmospheric pressure + water level pressure
2. If the gas pressure is lower than the atmospheric pressure, then the atmosphere pushes the water so that the water level in the sealed gas arm is higher than in the open arm. In this case, the gas pressure is computed as:
Gas pressure = atmospheric pressure - water level pressure
3. If the gas pressure is equal to the atmospheric pressure, then both gas and atmosphere exert an equal pressure on the water, making the water level in

both arms of the U-tube to remain the same. In this case:

Gas pressure = atmospheric pressure

The atmospheric pressure is usually regarded as about 760 mmHg, and the water level pressure is computed using the formula:

water level pressure = $h\rho g$

where h is the height difference between water levels in both arms of the U-tube as illustrated in Figure 4, ρ is the density of water (about 1000kgm^{-3}), and g is the value of acceleration due to gravity (about 10ms^{-2}).

As an illustration, let's have a look at the following question

If the liquid inside the U-tube of the manometer illustrated in Figure 4 is mercury and the height h is 25cm, determine the gas pressure.
(Take atmospheric pressure = 76 cmHg)

Solution!

Since the mercury level in the gas arm is higher than in the open arm, it implies that the gas pressure is less than the atmospheric pressure. And so:

Gas pressure = atmospheric pressure – mercury level pressure

Since the height difference in the mercury levels of both arms = 25cm, it implies that the mercury level pressure is 25 cmHg

Therefore:
Gas pressure = 76 cmHg – 25 cmHg
= 51 cmHg

Pascal's Principle

29

The principle which deals with the transmission of pressure in fluids is called Pascal's principle.

Pascal's principle states that the pressure applied to an enclosed fluid is transmitted undiminished to every portion of the fluid and the walls of the containing vessel.

The principle is only applicable to fluids (i.e. liquids and gases).

Applications of Pascal's Principle

30

The operation of instruments like, the car brake system, the siphon, the hydraulic press, the lift pump, the force pump, the bicycle pump and the syringe are all based on Pascal's principle. We are going to briefly discuss the operation of the car break, siphon, hydraulic press and lift pump in this book.

The car brake system

31

When the foot pedal of the brake is pressed, the piston in the master cylinder presses on the liquid. This pressure is transmitted to the wheel cylinder. The wheel cylinder has two opposed pistons which expand the brake-shoes which make contact with the wheel drum and brakes the car. On releasing the pressure on the pedal, the brake-shoe pull-off springs, force the wheel pistons back to the master cylinder. The pressure set up in the master cylinder is transmitted to all the four wheel cylinders, so all wheels are equally affected by braking.

Note!

32

No air should be allowed to enter the car braking system (i.e. the pipe containing the hydraulic fluid). This is because, unlike liquids, air can be compressed so with air present brakes will not work properly.

The Siphon

33

The siphon (illustrated in Figure 5) is a bent tube used for transferring liquid from a higher level to a lower level. The piece of flexible tube is filled with the liquid and one end (E) is kept below the surface of the liquid in the container from which the liquid is transferred. The liquid continues to flow out from E to C so long as C is lower. The Siphon work based on the principle that the initial pressure applied to pump the liquid from E to C is sustained by gravity.

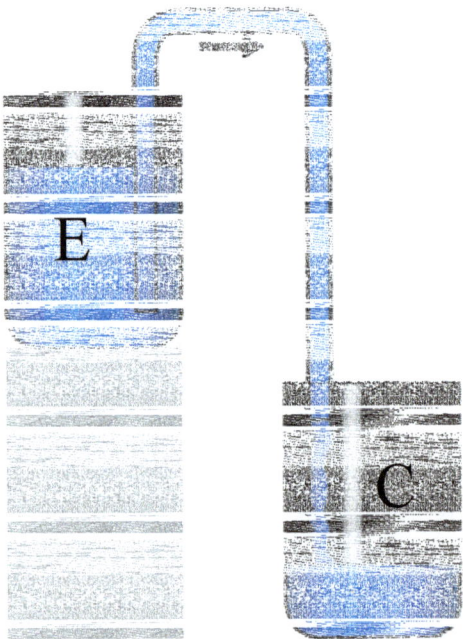

Figure 5. The Siphon (Wikipedia, 2017)

Note!

34

For the siphon to function efficiently;
- (i) The end C must be below the end E
- (ii) The tube must be full of the liquid; the presence of air in the tube can distort the flow
- (iii) The liquid must be raised a distance that is not greater than the barometric height for the liquid

The hydraulic press

35

The hydraulic press works based on the principle of transmission of liquid pressure. It is made up of two pistons; one piston of a small cross-sectional area, A_1, and the other piston of a large cross-sectional area, A_2. A liquid (usually special oil for the hydraulic system) is enclosed and used to fill the space between the two pistons.

Since the pressure in an enclosed liquid is transmitted undiminished to every portion of the fluid (according to Paschal's principle), a small effort (E) applied at A_1 will create a pressure that is transmitted undiminished to A_2 where it produces a magnified force. The force at A_2 is magnified because of the larger area. And so it can be used to lift a load which the initial effort could not lift.

Clearer explanation of the hydraulic press

36

Recall that pressure is $\dfrac{Force}{Area}$

So, if an effort E is applied at A_1, the pressure is:

$$\text{pressure} = \frac{E}{A_1}$$

and at A_2, the pressure that will be required to lift a load L is:

$$\text{pressure} = \frac{L}{A_2}$$

Now, by Pascal's principle, the pressure is transmitted undiminished from A_1 to A_2, so the pressure at A_1 is equal to the pressure at A_2. That is:

$$\frac{E}{A_1} = \frac{L}{A_2}$$

Therefore the force experienced at A_2 is:

$$L = \frac{A_2 E}{A_1}$$

And since A_2 is greater than A_1, it implies that L will always be greater than E.

Precisely, L will be 2 times greater than E if A_2 is 2 times greater than A_1, and L will be a million times greater than E if A_2 is a million times greater than A_1. The idea therefore is usually to make A_1 as small as possible, and A_2 as large as possible.

This is the whole idea behind using a small effort on the hydraulic press to lift very heavy cars and other loads.

The lift pump (common pump)

The lift pump or common pump (Figure 6) is used for raising water from wells. It consists of a piston working in a barrel or cylinder to which is attached the uptake pipe, which reaches down into the well of water. The piston head has a hole through it which can be closed by a valve A, while the second valve B closes the barrel at its base.

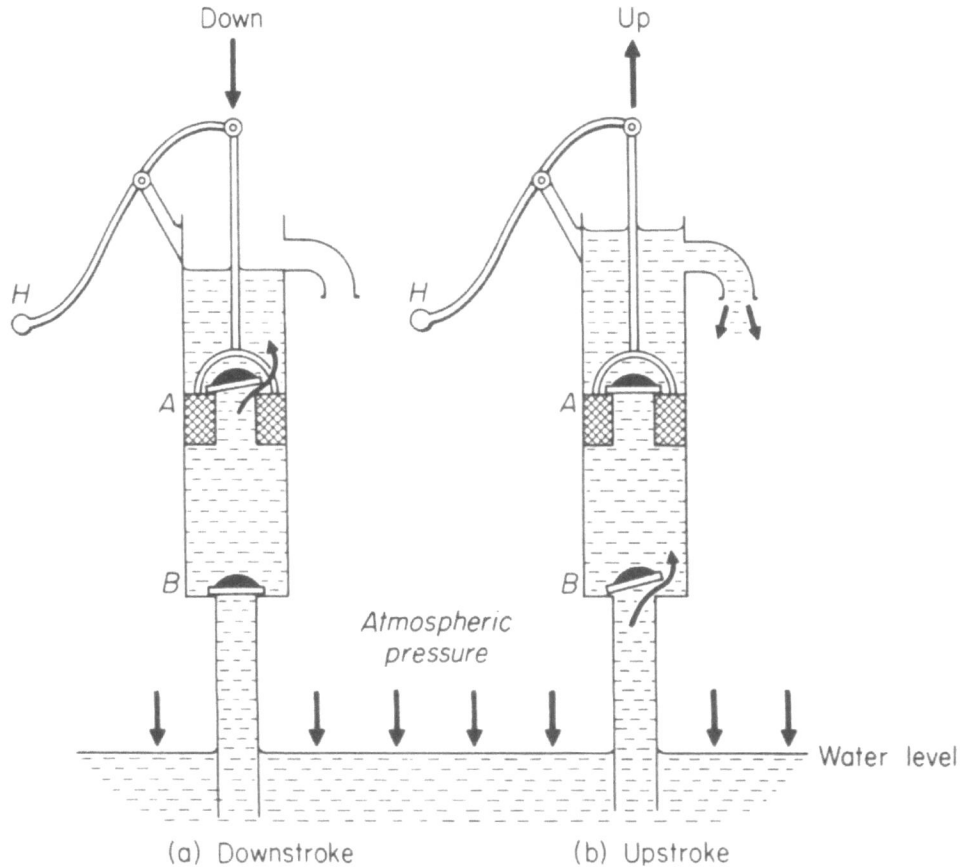

Figure 6. The Lift Pump (PhysicsMax, 2017)

Mechanism of Operation

During the upstroke, valve A is closed, and the pressure below the piston in the barrel falls. The atmospheric pressure on the surface of the water outside pushes the water, which opens valve B and rushes into the barrel. On the downstroke, valve B is closed, the water in the barrel opens value A and passes above the piston.

On the next upstroke, valve A closes and the water above the piston is raised until it flows out through the outlet pipe; valve B opens and as usual atmospheric pressure forces water into the barrel. And the cycle continues.

Note!

39

Since water comes into the barrel due to atmospheric pressure, and the column of water which the atmospheric pressure can support is about 10m. It means that the maximum height to which a lift pump can raise water is theoretically about 10m. In practice the height is usually less than 10m.

And that's it for the lift pump. If that's clear, we can round up with the following questions to test our understanding!

JAMB (1991)

40

Which of the following is most suitable for use as an altimeter?
(A) A mercury barometer   (B) A forte's barometer
(C) A mercury manometer   (D) An aneroid barometer.

Answer is (D) The Aneroid Barometer. See Plan 25 for explanation.

JAMB (2001)

41

The height at which the atmosphere ceases to exist is about 80km. If the atmospheric pressure on the ground level is 760mmHg, the pressure at a height of 20km above the ground level is

(A) 380mmHg   (B) 570mmHg   (C) 190mmHg   (D) 480mmHg

Solution

42

The information can be represented as shown below

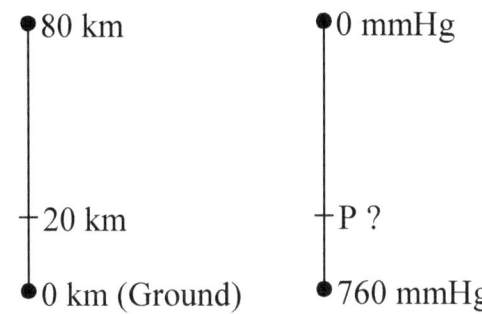

So by making a direct comparison, we get the equation below

$$\frac{P-760}{0-760} = \frac{20-0}{80-0}$$

$$\therefore \frac{P-760}{-760} = \frac{20}{80}$$

⇨ 80(P - 760) = -760(20)

   80P – 60800 = -15200

   80P = 60800 - 15200

$\therefore P = \dfrac{45600}{80} = 570$ mmHg

So option B is correct

And finally, we conclude with the following exercises!

Exercise

1. The hydrostatic blood pressure difference between the head and feet of a boy standing straight is $1.65\times10^4 Nm^{-2}$. Find the height of the boy. (Take Density of blood = $1.1\times10^3 kgm^{-3}$ and g = $10ms^{-2}$).

(A) 0.5m   (B) 0.6m   (C) 1.5m   (D) 2.0m

2. Which of the following statements are correct for the pressure in liquids?

(i) Pressure in a liquid at a point acts equally in all directions.

(ii) Pressure increases with depth.

(iii) Pressure of a depth depends on the shape of the container.

(iv) Pressure at the same depth in different liquids is proportional to the densities of the liquids.

(A) I, II and III only   (B) I, II and IV only

(C) I, III and IV only   (D) II, III and IV only

3. The atmospheric pressure at the top of water in an ocean is $1.3\times10^6 Nm^{-2}$. What is the total pressure at the bottom the ocean 10m deep? (Take density of water = $1000 kgm^{-3}$ and g = $10ms^{-2}$).

(A) $1\times10^7 Nm^{-2}$ (B) $1.4\times10^6 Nm^{-2}$ (C) $1.4\times10^5 Nm^{-2}$ (D) $1.0\times10^5 Nm^{-2}$

4. A pilot records the atmospheric pressure outside his plane as 63cm of Hg while a ground observer records a reading of 75cm of Hg for the atmospheric pressure on the ground. Assuming that the density of the atmosphere is constant. Calculate the height of the plane above the ground (relative density of Hg = 13.0 and that of air = 0.00013).

(A) 1,200m (B) 12,000m (C) 13,000m (D) 27,400m

5. A diver is 5.2m below the surface of water of density 1000kgm^{-3}. If the atmospheric pressure is 1.02×10^5Pa, calculate the pressure on the diver [g = 10ms^{-2}].

(A) 6.20×10^4Pa (B) 1.02×10^5Pa (C) 1.54×10^5Pa (D) 5.20×10^5Pa

6. When the Pressure of a fixed mass of a gas is doubled at constant temperature, the volume of the gas is

(A) Increased four times (B) Doubled (C) Unchanged (D) Halved

7. A reservoir 500m deep is filled with a fluid of density 850kgm^{-3}. If the atmospheric pressure at the top of the reservoir is 1.0×10^5Pa, calculate the pressure at the bottom of the reservoir (Take g = 10ms^{-2}).

(A) 5.25×10^6Nm^{-2} (B) 4.26×10^6Nm^{-2} (C) 4.35×10^6Nm^{-2} (D) 4.25×10^6Nm^{-2}

8. A reservoir is filled with a liquid of density 2000kgm^{-3}. Calculate the depth at which the pressure in the liquid will be equal to 9100Nm^{-2} (Take g = 10ms^{-2}, and ignore the atmospheric pressure).

(A) 26.2cm (B) 45.5cm (C) 66.4cm (D) 81.9cm

9. (I) Density of the liquid

(II) Acceleration due to gravity

(III) Type of container of the liquid

(IV) Shape of container of the liquid

Which of the above conditions will NOT affect the pressure in fluids?

(A) I and III only (B) II and III only (C) III and IV only (D) I and IV only

10. A rectangular block of dimensions 2.0m x 1.0m x 0.5m, weighs 400N. Calculate the maximum pressure exerted by the block on a horizontal floor.

(A) 100 Nm^{-2} (B) 2000 Nm^{-2} (C) 200 Nm^{-2} (D) 800 Nm^{-2}

Answers

1. C
2. B
3. B
4. B
5. C
6. D
7. C
8. B
9. C
10. D

References

Dojo University (2017). The manometer, http://learn.dojouniversity.com/wp-content/uploads/2017/01/standard-manometer.jpg

GlobalSpec (2011). Pressure in Liquids, http://www.globalspec.com/ImageRepository/LearnMore/201111/pressure

PhysicsMax (2017). The Lift Pump, https://physicsmax.com/the-common-pump-lift-pump-6225

SchoolPhysics (2016). The aneroid barometer, http://www.schoolphysics.co.uk/age1114/Mechanics/Statics/text/Barometer_aneroid/

Tutorvista (2017). The simple mercury barometer, http://image.tutorvista.com/content/matter-states/mercury-barometer.jpeg

Wikipedia (2017). The Siphon, https://en.wikipedia.org/wiki/Siphon

www.ingramcontent.com/pod-product-compliance
Lightning Source LLC
Chambersburg PA
CBHW051834210526
45473CB00005B/1871